THE ATMOSPHERE:
Five Billion Million Tons of Air

(Original French title:
L'atmosphère)

by Jean-Pierre Maury

*Translated from the French by
Albert V. Carozzi and Marguerite Carozzi*

New York • London • Toronto • Sydney

First English language edition published in 1989 by
Barron's Educational Series, Inc.

© 1987 Hachette/Fondation Diderot-La Nouvelle Encyclopédie, Paris, France.

The title of the French edition is *L'atmosphère*

All inquiries should be addressed to:
Barron's Educational Series, Inc.
250 Wireless Boulevard
Hauppauge, NY 11788

International Standard Book No. 0-8120-4211-5

Library of Congress Catalog Card No.   89-6899

Library of Congress Cataloging-in-Publication Data

Maury, Jean-Pierre
    [Atmosphère, English]
    The Atmosphere: five billion million tons of air/by Jean-Pierre Maury:
translated from the French by Albert V. Carozzi and Marguerite Carozzi.
    p.   cm.—(Barron's focus on science)
    Translation of *L'Atmosphère*.
    Includes index.
    Summary: Describes the composition and characteristics of the thin ocean
of air that surrounds the earth explaining its function and interrelationship
with living things.
    ISBN 0-8120-4211-5
    1. Atmospheric pressure—Juvenile literature. [1. Atmosphere.
2. Atmospheric pressure.] I. Title. II. Series.
QC885. M2913 1989
551.5'4--dc20   89-6899
                                                     CIP
                                                     AC

PRINTED IN FRANCE
901   9687   987654321

CONTENTS

The Fear of Vacuum

For over 2000 years, humans have known how to make and operate pumps to raise water. It has been less than 400 years, however, since they understood how these pumps work.

Indeed, for a long time their functions were "explained" by saying that "Nature abhors a vacuum." It was believed that using a pump created a vacuum in a pipe above the water, and since Nature abhorred this vacuum, it hastened to fill it with water, which thus rose in the pipe.

T his was the explanation accepted from the time of Aristotle (fourth century B.C.) until the seventeenth century.

Ten Meters (Thirty-three Feet) but No More

Nevertheless, workers who installed pumps knew very well that the pumps worked only if the water table was not too far below the surface. If the water level at the bottom of the well was below 10 meters (33 feet) in depth, the pump was not able to raise the water to the surface.

However, although workers knew about the problem, naturalists appeared to ignore it completely because their science had little to do with *everyday reality*.

This attitude changed at the beginning of the seventeenth century with Galileo. He believed that science must take care of *everyday reality* and that the role of science consists precisely in explaining what really happens. Ever since 1590, he had undertaken numerous experiments on the fall of bodies, on movement, and also on floating bodies. After building his telescope in 1609, he be-

came particularly interested in astronomy. However, in 1633, a decree of the Church forbade him to work in that field and he thus returned full-time to experiments in physics.

In 1640, Galileo was the scientific expert to the Grand Duke of Tuscany, who had decided to build fountains in his garden by pumping water from aquifers located at 13 meters (42.6 feet) depth. When it was noticed that pumps were unable to bring this water to the surface, Galileo was called in.

Galileo made measurements and found that water rose in

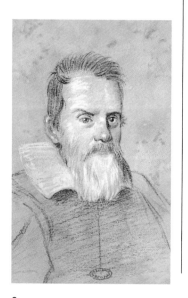

pipes to a height of about 10 meters (33 feet) and then stopped. Further pumping was of no use: the water rose no higher. Nevertheless, when pumping, a vacuum is created above the water, a vacuum that Nature should abhor: what happened?

Galileo did not find the solution, but even though very old and ailing, he thought about it until his death in 1642. Only two years later, one of his students, Torricelli, finally explained the working of pumps.

The Weight of Air

For Torricelli, it was not a "fear of vacuum" that caused water to rise in the pipe of a pump, it was the weight of the air that pressed upon the surface of the water all around the pipe but not against the inside of the pipe since air had been removed.

Similarly, if a hand presses on a ball of soft dough while spreading the fingers a little, the dough rises between the fingers only in those places where the hand is not pressing.

This is why water stops rising when it has reached a certain height above the surface: at that moment the weight of the water column is just strong enough to replace the weight of air that would press there if the pipe were open at its top.

How could Torricelli prove that his was the correct explanation? He had the idea to use a liquid 13.6 times heavier than water—mercury. According to him, a column of mercury of the same weight as a water column should therefore rise, in the same tube, 13.6 times less high. Furthermore, if the weight of the air supports a water column of 10 meters (33 feet), it should support a column of mercury of 76 centimeters (30 inches.)

For this experiment, a pump was not even necessary! Torricelli used a 1 meter (3.3 feet) long tube of glass, closed at one end and filled with mercury. He sealed it with his finger, turned it, and plunged its open end into a dish of mercury.

The mercury in the tube dropped a little and stopped when its column was 76 centimeters (30 inches) high. In the upper part of the tube, above the mercury, was a vacuum, exactly as if one had used a pump. Up there, nothing pressed against the small surface of mercury. On the contrary, air pressed quite

The level of mercury in Torricelli's tube is 76 centimeters (30 inches) above its level in the dish.

strongly upon the mercury in the dish all around the tube and supported a column of mercury measuring 76 centimeters (30 inches) in height, but no more. Indeed, this column had the same weight as a 10 meters (33 feet) high water column of the same diameter. Torricelli was now quite certain that at last he had found the correct explanation.

Even though he was convinced, however, other naturalists were not because one does not give up easily such an old notion as fear of vacuum. During many months, a violent controversy spread over Europe.

The man who finally put an end to the controversy and buried the fear of vacuum forever was a young French naturalist, Blaise Pascal.

The Use of Mountains

Pascal repeated Torricelli's experiment and invented many others to show that Torricelli was correct. However, the most brilliant of all experiments, the one that finally convinced everybody, required a mountain for its undertaking!

Why a mountain? Because there is less air over its top so that it should press less than at sea level. If Torricelli's experiment were done on top of a mountain, and if Torricelli was indeed correct, then mercury should rise less than 76 centimeters (30 inches). If, on the other hand, it is the force of vacuum, supposedly abhorred by Nature, that makes mercury rise—as many naturalists continued to believe—the column would have the same height on top and at the foot of a mountain since there is no reason that this force of vacuum should be less strong on top of a mountain than in a valley.

Pascal's brother-in-law lived at Clermont-Ferrand, near the Puy-de-Dôme. Asked by Pascal, he made the experiment in 1647 on top of that mountain and found that the column of mercury indeed measured only 65 centimeters (25.5 inches) instead of 76 centimeters (30 inches). The experiment provided such a clear result that Pascal himself could repeat it in Paris, in front of a large audience, using the top and the base of the tower of Saint-Jacques.

To honor this historic experiment, a statue of Pascal was placed at the foot of the tower.

Teams of Horses at Magdeburg

Soon machines that were a kind of pump without water were constructed that could produce a vacuum in a container. These vacuum pumps allowed interesting experiments.

In 1650, Otto von Guericke, mayor of the German city of Magdeburg, ordered two hollow hemispheres of bronze to be cast. When these two parts were joined to form a complete sphere (80 centimeters [31.5 inches] in diameter), and when the volume inside had been evacuated, the pressure of the outside air was so strong against the two halves that two teams of eight horses could not separate them!

The atmosphere is thus capable of exerting enormous forces. How could these forces be put to use? In fact, in the past, other than the muscles of people and horses, small waterfalls that were used to turn a mill, or the wind that pushed boats and turned the wings of a windmill, no other motor existed. However, to use the forces of the atmosphere, a vacuum— at least a partial one—had to be made in a container. Machines

consisting of a cylinder and a piston were invented. Once vacuum was made in the cylinder, and as soon as the piston was released, the atmosphere would exert a force upon it, a force that could be used to lift weights, to pump water, and so on.

At first, however, the cylinder had to be evacuated. In 1690 Papin had the idea of using the cooling of water vapor for that purpose, and he thus built the first ancestor of the steam engine.

Much earlier, however, in 1648, Pascal succeeded in calculating the total weight of the atmosphere.

How Heavy Is the Entire Atmosphere?

In an evacuated tube, water rises 10 meters (33 feet). If this tube has a section of 1 square centimeter (.155 square inch), it thus holds 1000 cubic centimeters (61 cubic inches) of water, that is, 1 liter (1 quart) or a kilogram (2.2 pounds) of water.

In the same tube, mercury would rise 76 centimeters (30 inches) and it would hold a column of mercury weighing 1 kilogram (2.2 pounds).

If the tube were open at the top, what would replace water or mercury? Air, of course! In other words, a column of air above a surface of 1 square centimeter (.155 square inch) weighs 1 kilogram (2.2 pounds).

Therefore, there is 1 kilogram (2.2 pounds) of air above every square centimeter (.155 square inch) of the surface of the Earth. To know the total mass of the atmosphere in kilograms, it is necessary only to measure the surface of the Earth in square centimeters.

The surface of the Earth measures 5 billion billion square centimeters. Therefore, the atmosphere weighs 5 billion billion kilograms, that is, 5 billion million tons.

This result was reached by Pascal as early as 1648 (in other units: the weight was in pounds and the surface in square feet, and so on).

Drinking with a Straw

When sucking on a straw, one does the same thing as a pump: one removes the air that filled the straw. At that place, air exerts less pressure upon the surface of the liquid whereas it presses as strongly as before upon the surface around the straw. Because of these forces, the liquid rises in the straw since it is the only place where nothing hinders this movement. It is the weight of air around the straw that pushes the liquid up toward the mouth.

The Ears of Divers

When we dive only a little below the surface of water, our ears hurt. This is, by the way, not the only time our ears hurt. It also happens in a cable car, in a small plane, or even in a fast elevator in a tall building, in short, every time one changes rapidly either elevation in the air or depth in water. However, in water it suffices to dive only 1 meter (3.3 feet) to start feeling pain in our ears whereas in the air it is necessary to fall or rise rapidly some hundred meters (328 feet) at least. Therefore, we shall first study this effect in water.

On Every Square Centimeter (.155 Square Inch)

As we dive in water, the water presses more and more upon our eardrums, and elsewhere also of course: it presses equally hard on our forehead or on our shoulders, but there we do not feel the pressure.

On every square centimeter (.155 square inch) of skin, water presses with a certain force. If this force is, for example, equal to the weight of an object weighing 1 1/2 kilograms (3.3 pounds), it is said that the "water pressure" on the skin at that place is 1.5 kilograms/square centimeter (3.3 pounds/.155 square inch).

Similarly, each time a certain force is spread over a certain surface, the force divided by the surface is called the pressure.

For instance, a man weighing 80 kilograms (176 pounds) who steps on the ground on a surface (his heels and soles) of 200 square centimeters (31 square inches) exerts a pressure on the ground of 80/200, that is, 0.4 kilograms/square centimeter.

The same man with larger soles of course has the same weight; however, his weight is spread over a larger surface and hence the pressure exerted on the ground is lessened. This is why snowshoes allow us to walk on soft snow. They have a surface ten times larger than that of ordinary soles and hence the pressure exerted on the ground is ten times less than that of ordinary shoes. This time, the pressure is 0.04 kilograms/square centimeter, for example, light enough for a man to walk on snow without sinking into it up to his knees.

In contrast, a woman wearing spike heels steps on a very small surface and exerts a very strong pressure on the ground: her heels produce holes in a parquet floor! In fact, let us imagine a woman weighing 50 kilograms (110 pounds) with heels of 1 square centimeter (.155 square inches). When her body rests on one heel alone, that heel exerts a pressure of 50 kilograms/square centimeter on the floor!

By spreading a mountain climber's weight and knapsack over a large surface, snowshoes reduce the pressure she exerts on the snow. She walks without sinking, even if the snow is soft.

If the Diver Turns His Head

Water presses on the eardrums of a diver with a certain force and hence exerts a certain pressure on them. What does this pressure depend upon?

Upon depth of course: pain increases rapidly when diving under water. This can be explained easily: the deeper we dive, the more water exists above that presses on us.

However, there is one factor upon which the pressure of the water does not depend: direction.

If the diver turns his head, the pain does not change as long as depth remains unchanged. If his ear is turned toward the surface, water presses downward. If it is turned toward the bottom, water presses upward with the same pressure (if the depth remains the same).

If the ear is vertical, water presses sidewise, but always with the same pressure.

One can thus talk about *pressure* at a certain depth of water. For example, at 10 meters (33 feet) depth, the pressure is 2 kilograms/square centimeter (4.4 pounds/.155 square inch). This means that if a diver is at a depth of 10 meters (33 feet), water presses on each square centimeter (.155 square inch) of his body with a force equal to the weight of an object of 2 kilograms (4.4 pounds).

The Diver on the Beach

Once she has left the water and is stretched out in the sun on the beach, is the diver no longer subject to pressure? Yes she is, to that of the air!

Torricelli has shown that on a horizontal surface of 1 square centimeter (.155 square inch), air presses with the same force as a weight of 1 kilo (2.2 pounds). However, this is not true only for a horizontal surface: both in the air and in the water, one can talk of *pressure* at any given place. This pressure is 1 kilogram/square centimeter (2.2 pounds/.155 square inch) at sea level.

Since this pressure at sea level is exerted by the atmosphere, it has been given the name "1 atmosphere." One atmosphere is thus 1 kilogram/square centimeter. Any pressure can thus be measured by atmospheres: at 10 meters (33 feet) water depth, pressure equals 2 atmospheres.

This means that at sea level air presses with a force equal to a weight of 1 kilogram (2.2 pounds) on any 1 square centimeter (.155 square inch) of surface, whether horizontal or not. This force is always perpendicular to the surface: it presses down vertically like a nail that one would wish to drive into a board.

One can thus sketch, for example, some of the forces of pressure exerted by atmospheric air on a round balloon. Notice that on top of the balloon the air presses downward, whereas it presses upward on the underside of the balloon. It is easy to figure out why this is necessary: look at what happens to a sheet of paper.

If a sheet of paper is held horizontally, the air presses on its surface with an enormous force. For an ordinary sheet with a surface of 600 square centimeters (93 square inches), this force is equal to the weight of an object weighing 600 kilograms, (1320 pounds), such as a cow!

It is obvious that the sheet itself does not support this weight! What supports it is a force equally great exerted by the air on the underside of the sheet and thus upward.

A very simple experiment can be done to prove this. A glass is filled completely with water and a sheet of paper, a little larger than the glass, is placed on top. The glass is then

turned while the sheet of paper is kept in place with one hand. Thereafter, with the other hand one continues to hold the glass while the hand that held the paper is removed: nothing happens! Nevertheless, it is advisable to run this experiment above the sink.

What holds the water inside the glass is evidently the pressure exerted by atmospheric air underneath the sheet of paper.

A Difference in Pressure

Let us return to our sheet of paper held by one hand. If the air did not press exactly as much on both sides, the sheet would not remain in place for a second. Indeed, the smallest *difference in pressure* between the two sides would submit the sheet to an important force. This is also what happens to a diver's eardrums when he plunges into the water.

The eardrum, which seals the bottom of the ear, is a skin stretched over a space filled with air, called the middle ear. This area communicates with the pharynx by a very slender tube, the Eustachian tube. Normally, the pressure of air is the same on both sides of the eardrum because the inner air communicates with the outer air through the Eustachian tube.

However, the Eustachian tubes are very thin and sometimes they tighten and become completely closed. The inner air is then separated from the outer air.

As long as the external pressure does not change, there is no force exerted on the eardrum. As soon as one rises in a cable car, for example, or a small plane, the pressure of the outer air decreases. This was shown

for the first time by Pascal's experiment on the Puy-de-Dôme. Thereafter, the pressure is no longer the same on both sides of the eardrum: the inner air presses more than the outer air and one starts to feel pain.

To stop this pain, the pressure on both sides of the eardrum must be equalized; that is, the inner air must communicate with the outer air. For this purpose, the Eustachian tube must be opened. This may be achieved by swallowing, which moves the muscles. Sucking candy helps because swallowing is made easier.

This is why candy is sometimes distributed to passengers of small planes that are not "pressurized." Larger planes flying at 10,000 meters (39,370 feet) altitude are pressurized, which means that there is no communication at all between the inside and the outside. This allows a pressure to be maintained in the cabin that is much higher than that of the outside air. In this case, a difference in pressure exists not between the two sides of an eardrum but between the two sides of a window in the plane. At 10,000 meters (39,370 feet) altitude,

The windows of pressurized planes are small because the difference in pressure between the inside and the outside air is great, as is the force exerted on every square inch of the window.

the outside pressure equals about 0.2 atm. Inside the plane, a pressure of 0.8 atm is maintained, which is a little lower than at sea level. There is therefore a difference in pressure of 0.6 atm between the inside and the outside. On a plane window of 600 square centimeters (93 square inches), we have a difference of 360 kilograms (792 pounds) between the force that presses toward the outside and the one that presses toward the inside. These windows must be very sturdy and not too large.

This is also the reason a pressure of 0.8 atm, instead of 1 atm, is maintained inside the plane. This reduces the difference with the outside so that the walls of the plane are less exposed to damage. It also means that the passengers must suffer during takeoff and landing.

It is the opposite with the diver: when he dives, the pressure is stronger outside his eardrums than in the "boxes" they cover. Therefore, extra air must enter these boxes. This is difficult because the Eustachian tubes do not open easily in that direction. It is not enough to swallow; one must blow very hard while closing the mouth and pinching the nose.

When the Diver Descends 10 Meters (33 Feet)

The deeper one dives into water, the stronger the pressure because above us is an increasingly higher water column. Is it possible to know the pressure at a given depth? Very easily.

Let us remember pumps for raising water and Torricelli's explanation. Atmospheric pressure supports no more than a water column of 10 meters (33 feet). This means that the pressure equals 1 atm at the bottom of a water column of 10 meters (33 feet) *overlain by the vacuum produced by the pump*.

Let us now imagine a water column of 10 meters (33 feet) in a tube open at the top. At the top of the water column, the pressure now equals 1 atm: the top of the column supports the entire column of air above it. And the bottom of the water column? It must support the water column *plus* the overlying air column.

The same thing happens when acrobats perform a "human pyramid." A very strong man carries on his shoulders a woman who in turn carries a little girl on her shoulders.

Space Vacuum

Satellites and spacecraft operate in an area above an altitude of 100 kilometers (62 miles), where the atmospheric pressure is practically zero. At such an altitude the air is so "rare" that one talks of "space vacuum." This vacuum surrounds all spacecraft.

Those flown by people are pressurized, as are planes flying at high altitudes. A pressure of about 0.8 atm is maintained, which is sufficient for the breathing and comfort of astronauts. When they walk in space, they slip into airtight, pressurized spacesuits.

One sometimes has the impression that pressurization in spacecraft poses more problems than in planes because of the almost total vacuum in space. Well, it does not make much difference. If the outside pressure is zero instead of 0.2 atm as at 10,000 meters (39,370 feet) altitude, the difference between inside and outside is only 0.8 instead of 0.6 atm. The windows must be a little sturdier, but not much.

The man's shoulders support the weight of the woman plus the weight of the little girl.

The woman supports the little girl, whereas the man supports the woman *plus* the little girl.

It is the same for water. The woman's shoulders represent the top of the water column upon which air presses. The man's shoulders represent the bottom of the column on which rest the water column *plus* the air. If the water column measures 10 meters (33 feet), the pressure at the bottom is 1 atm *plus* 1 atm, that is, 2 atm. Hence at a 10 meter (33 feet) depth of water, the pressure is 2 atm.

And farther down? For each additional 10 meters (33 feet) of water, 1 atm is added. This gives the following scale:

At a depth of 0 meters (0 feet):	1 atm
At a depth of 10 meters (33 feet):	2 atm
At a depth of 20 meters (65.6 feet):	3 atm
At a depth of 100 meters (328 feet):	11 atm
At a depth of 1,000 meters (3,280 ft):	101 atm
	and so forth.

Of course, it is not necessary to descend by steps of 10 meters

(33 feet) at a time. At 5 meters (16 feet) depth, the pressure is 1.5 atm; at 2 meters (6.5 feet) depth, 1.2 atm, and so forth.

Therefore, if a diver is at 2 meters (6.5 feet) depth, the water pressure is already 1.2 atm. If he has not already "unblocked his ears," the pressure under his eardrums is the same as before diving: 1 atm. Therefore, there is a difference of 0.2 atm between the two sides of the eardrums, which causes some pain.

The Lock

The water in a canal has a perfectly horizontal surface, whereas the ground itself does not. If one wants to have a canal climb a hill, it must be separated into several sections in which water is at different levels like the steps of a "water staircase."

To go from one step to the other, a lock is necessary.

A lock is a watertight chamber that, by means of gates at either end, provides communication between two parts of a canal at different levels. The upstream gate separates the chamber from the upper part of the canal. The downstream gate separates the chamber from the lower part. Furthermore, at the bottom of these gates are small openings, called sluices, which can be opened or closed as necessary.

Suppose that the lock is full of water: the water level inside the chamber is the same as in the upper part of the canal. The two gates and their sluices are closed. What is the difference in pressure between the two sides of a sluice of the downstream gate?

If there are 2 meters (6.5 feet) more water inside, the difference of pressure between the two sides, near the bottom, is 0.2 atm. Hence water presses much more from the inside than from the outside. If a sluice is opened, the inside water rushes toward the outside until the water level is the same on both sides of the door: the lock is being emptied.

Once the water level is the same on both sides of the downstream gate, there is no longer a difference in pressure between the two sides and the gate can be opened. A barge coming from the lower part may enter the lock. The downstream gate is then closed and the sluices of the upstream gate are opened: water rushes into the lock and fills it until the water level is the

same as in the upper part of the canal. At that time, the upstream gate can be opened and the barge can continue its trip. It has "climbed a step" in the canal.

The Dam

Here, too, the difference in water level between the two sides of the dam produces a difference in pressure. For example, if there is 10 meters (33 feet) more water upstream, the pressure near the bottom on that side is 1 atm more than on the other. If an opening is made at the bottom of the dam, water rushes into it, and if a turbine (a kind of improved propeller) is placed in that hole, water makes this turbine turn and press very strongly against its blades. The turbine may thus turn an alternator to produce electricity (like the dynamo of a bicycle).

The greater the difference in the water levels on both sides, the greater the difference in pressure. This is achieved by penstocks, large pipes that carry water to the turbines from a dam located 100 or 200 meters (328 or 656 feet) higher up. This time, water presses with 10 or 20 atm on the turbine and also against the walls of the pipes, at least in the lower parts. Therefore, these pipes must be very sturdy!

Even an ordinary dam must be very sturdy. From beneath the upstream water level, the difference in pressure between the two sides of the dam increases downward until the downstream water level is reached. Therefore, a dam must be much thicker at the base than at the top.

Furthermore, dams are often curved (seen from above). The curved side is pointing toward the highest water level. Thus, the forces of pressure tend to compress the dam and wedge it strongly against the rocks on both sides: thus a dam is built that is very sturdy and requires less concrete.

Similarly, the gates of locks always open upstream. When closed, a pair of gates meets at an obtuse angle pointing upstream to withstand the water pressure. Thus, the forces of pressure always have a tendency to close the gates even tighter instead of opening them as they would if the obtuse angle were facing downstream.

Finally, another type of dam,

on a smaller scale, presents the same problems: the wall of an aquarium. For a water height of 50 cm (19.6 in) the difference in pressure between the inside and the outside, near the bottom, is 0.05 atm. Upon every square centimeter (.155 square inch) presses a force equal to an object of 50 grams (1.6 ounces) weight, which does not seem enormous. Nevertheless, on a horizontal band 5 cm (1.9 in) high and 80 cm (31 in) wide, this represents a force equal to a weight of 20 kg (44 lb), which is not negligible. Therefore, rather thick glass is used, otherwise all of the water—and its inhabitants—run the risk of finding themselves on the floor.

The Water Tower

If we open a faucet on the third floor, water starts to flow. It comes from pipes located in the ground. What causes the water to rise to the third floor?

In fact, it would be better to say "to rise again." Indeed, buried pipes that carry water to a house come from a reservoir located on a nearby hill, or from the top of a tower called a water tower.

Suppose that the water level in the reservoir is 20 m (65.6 feet) above the pipes. What is the water pressure in the pipes? It is the atmospheric pressure plus 20 m (65 ft) water, a total of 3 atm.

If the faucet on the third floor is 10 m (33 ft) above the pipe, water pressure in the faucet is that of the pipe *minus* 10 m (33 ft) water: 3 atm *minus* 1 atm = 2 atm. Hence, if the faucet is opened, water may spurt out violently as every child knows.

The water in park fountains does not really rise either: it rises again. It also comes from a reservoir or water tower located higher than the top of the fountain.

How does water rise from pipes buried in the ground to a faucet in a bathroom on the third floor?

In reality, it does not rise. It rises again because it comes from a water tower that is higher than the bathroom. It actually flows down from the water tower to the bathroom.

In a high-rise of forty floors, it would be different. To bring water to floors higher than the water tower, a system of pumps driven by electric motors would be necessary.

Archimedes' Discovery

Everybody has heard the story about Archimedes leaving his bathtub and shouting Eureka (I found it!). This happened about 2200 years ago and it is difficult to know if it is true. At any rate, whether in his bathtub or not, Archimedes really discovered something, and that something was very important: "Archimedes' principle."

What does this principle say? "A body (completely or partially) submerged in a fluid is buoyed up by a force equal to the weight of the fluid it displaces."

This allows us to understand quite a lot of things: why boats float, why water supports a swimmer provided he or she is well submerged, and so on. We can do even better: we can trace the origin of Archimedes' principle and understand why it is true from what we have previously learned about pressure in water.

The Difference Between Up and Down

Let us imagine a body that is not moving in water, for instance a diver who examines a wreck found at the bottom of the sea.

Water presses on her body from all sides as well as on her air container, mask, and so on. Water presses downward on her back. On her stomach, it presses upward. Will all these forces cancel each other? No! The diver's stomach is at greater depth than her back so that

water pressure is stronger there and hence water pushes more strongly upward against her stomach than downward on her back. As a result, the difference between these two forces is an upward or buoyant force according to Archimedes' principle.

It is even possible to know exactly the magnitude of this force. Let us imagine that we replace the diver by a kind of extremely thin shell that has exactly the shape of the diver's body and is filled with water. On the stomach of this "water diver," the forces of pressure are the same as on the stomach of the real diver since the position and the location are the same. This also applies to the back. The difference, that is, the buoyant force, is exactly the same for the "water diver" as for the real diver.

This difference can be easily measured for the water diver. Consisting of water, it floats perfectly without tending to rise or sink. The force that supports it—the buoyant force or buoyancy—is thus exactly equal to its weight. This force is the same for the real diver. Buoyancy for the real diver is equal to the weight of the "water diver," that

is, to the weight of the water that exactly occupies the volume of the diver. According to Archimedes' principle, this is the weight of the displaced fluid, displaced because the diver has taken its place.

What Carries the Bottles of Air?

To repeat, the buoyant force on a submerged body in water is equal to the weight of the water that would occupy its place. Since water weighs 1 kg/liter (2.2 pounds/quart), this force often becomes very important.

For instance, the submerged bottles that hold compressed air, necessary for the diver's breathing, are large cylinders made of thick steel. They are very heavy, noticed when the divers lift them to put them on their backs.

In water, however, a child can hold them with a finger! Indeed, if a bottle of 8 kg (17.6 lb) has a volume of 8 liters (8.5 quarts), water exerts a buoyant force equal to the weight of 8 liters (8.5 qts) water, that is, 8 kg (17.6 lbs). Hence, the bottle weighs "nothing at all" in water! At least, it weighs nothing for the diver because water, not

the diver, carries it!

And the diver? He weighs "less than nothing!" In fact, his body is only a little bit heavier than the "water diver;" that is, his weight is a little less than the buoyant force that pushes him upward. This is why it is possible to float on one's back.

Floating on One's Back

This means floating on the surface of the water without moving. In this position, the buoyant force supports the body perfectly because the force is equal to the weight of the body. Actually, the swimmer is not completely submerged or she would not be able to float on her back for long! In seawater, half her head should be above the surface of the water, and a little less in freshwater because it is lighter. The buoyant force on the rest of the body is sufficient to support all the weight.

When a swimmer is completely submerged, the buoyant force is thus a little greater than her weight, and hence a "water diver" is heavier than the real swimmer. The human body is indeed a little lighter than water.

This factor depends of course upon individual people. Fat is lighter than muscles, which in turn are lighter than bones. Women, for instance, have lighter bones than men in proportion to their size and hence they float better. A skinny old man with heavy bones and not much else would barely float.

Since the human body is normally lighter than water, one wonders how anybody can drown. As a matter of fact, although the human body is light, the stored air in the lungs plays a critical role. Indeed, an endangered swimmer who panics is unable to keep his mouth above water and ends up "breathing water." Thereafter he does not float—he is heavier than water, and he sinks.

Loading a Barge

When empty, a barge rides very high. When full, it rides very low. Its weight having increased, the buoyant force that supports it must also increase. The only possible way is that the displaced volume of water—the volume of water displaced by the barge—also increased.

It is even possible to use

Archimedes' principle to calculate the load of a barge if one knows its size, regardless of the nature of its load (sand, coal, or pillows). Let us imagine a more or less rectangular barge, 15 meters (49 feet) long and 3 meters (9.8 feet) wide, which sinks 1.5 meters (7.4 feet) when loaded.

The volume of water displaced by the barge has thus increased 15 x 3 x 1.5, that is, 67.5 cubic meters (88.3 cubic yards). The buoyant force has increased by a value equal to the weight of 67.5 cubic meters (88.3 cubic yards) of water, that is, 67.5 tons. The load of the barge is thus 67.5 tons.

Floating Ice

Ice cubes float in a glass of water. As for the swimmer who floats on her back, this means that to support the weight of the ice cube it suffices to "displace" a volume of water a little smaller than that of an ice cube. Consequently, an ice cube weighs less than a volume of water equal to its own volume. In other words, 1 liter (1 quart) of ice weighs less than 1 liter (1 quart) of water.

In fact, 1 liter (1 quart) of water weighs 1 kilogram (2.2 pounds), whereas 1 liter (1 quart) of ice weighs only 0.9 kilogram (1.9 pound). To support an ice cube of 1 liter (1 quart), it is thus necessary to displace a volume of water that weighs 0.9 kilogram (1.9 pound), that is, 0.9 liter (quart) water. An ice cube floats with nine-tenths of its volume under water and only one-tenth above the surface.

Huge ice cubes float in the polar seas that are called icebergs (from a Norwegian word meaning "mountain of ice"). They are very dangerous to ships. Everybody knows about the *Titanic*, which collided with an iceberg in April 1912 and sank. Part of the danger represented by an iceberg is of course that only one-tenth of its volume is above the water surface. The "mountain of ice" visible is only part of a mountain that is nine times larger, the shape and position of which remain unknown!

In regard to floating ice, try this simple experiment:

A glass of water with an ice cube in it is filled to the rim. Part of the ice cube stands out and is higher than the rim. Does the water overflow when the ice cube melts?

Not by a droplet! In fact, the volume of water displaced by the ice cube has the same weight as the ice cube itself and hence the same weight as the water produced by its melting. This water therefore occupies exactly the volume displaced by the ice cube without the smallest droplet flowing over the rim!

The Buoyant Force of the Air

Air is of course much lighter than water but it nevertheless has a considerable weight. The air surrounding us weighs close to 1.3 grams/liter (0.5 ounces/quart) at sea level. This is not a lot, but 1 cubic meter (1.3 cubic yards) equals 1000 liters (900 quarts) so that 1 cubic meter (1.3 cubic yards) air weighs 1.3 kilograms (2.9 pounds). A normal sized room of 15 square meters (161 square feet) that is 2.7 meters (8.8 feet) high holds about 40 cubic meters (52 cubic yards), that is, more than 50 kilograms (110 pounds) of air! However, a theater hall that is 20 meters (65.6 feet) long, 10

If the audience in a theater hall were asked how much air the hall holds, how many would answer several tons?

meters (33 feet) wide, and 10 meters (33 feet) high has a volume of 2000 cubic meters (2615 cubic yards) and thus holds about 2.6 tons of air.

Being heavy, the air presses on objects emerged in it with a vertical buoyant force equal to the weight of the displaced air, that is, the weight of air that would occupy the volume of the object if that object were not there.

How much is this force on a human? A man's volume equals about 75 liters (67.5 quarts). He thus displaces 75 liters (67.5 quarts) air, which weighs about 100 grams (3.5 ounces). The air exerts a buoyant force on the man equal to the weight of an object of about 100 grams (3.5 ounces). He does not risk flying away. However, an object as large as the man that weighs less than 100 grams (3.5 ounces) would fly away because the buoyant force on this object would be greater than its weight. What kind of object weighs less than 100 grams (3.5 ounces) and has the volume of a man? A light balloon, inflated with gas lighter than air.

The lightest gas is hydrogen; it is 14 times lighter than air. We have learned that 75 liters (67.5 quarts) air weighs about 100 grams (3.5 ounces); 75 liters (67.5 quarts) hydrogen would thus weigh 14 times less, or about 7 grams (.25 ounce). If these 7 grams (.25 ounce) locked up in a balloon that weighs 50 grams (1.75 ounce), the inflated balloon would weigh 57 grams (2 ounces).

A balloon of 75 liters (67.5 quarts) inflated with hydrogen is therefore pushed downward by the weight of an object of 57 grams (2 ounces) and lifted by a buoyant force equal to 100 grams (3.5 ounces): the balloon flies away. If held by a string, it stretches with a force equal to the difference between the buoyant force and its weight, that is, equal to the weight of an object of 43 grams (1.5 ounces) (the difference between 100 and 57).

If, instead of holding the balloon, an object weighing less than 43 grams (1.5 ounces) is attached, the balloon flies away and carries along the object.

Once it has flown away, what happens to the balloon?

When Balloons Rise

Sometimes it explodes, which is easily explained. When the

balloon rises into the air, the external air pressure decreases and hence the difference in pressure between the inside and the outside of the balloon increases. The same thing happens to the balloon if one tries to inflate it too much on the ground. If it is fragile and already fully inflated, it explodes.

If it were not fully inflated when taking off, its volume would increase as the external pressure decreases. This happens to weather balloons, which rise very high.

And if the balloon is already fully inflated when departing but very sturdy? It will rise without exploding but less and less fast. When it reaches a certain altitude, it will stop rising.

If the wind is blowing, the balloon moves together with the surrounding air and thus travels great distances, always remaining at the same height. Then little by little it lets air leak out, and finally it falls back to the ground. If a constant wind is blowing, balloons can sometimes fly hundreds of miles without going above a certain altitude.

Why does a balloon stop rising when it has reached a certain altitude? The answer is easily guessed.

When it stops and neither rises nor sinks, the upward force is equal to its weight. This weight has not changed since it left the ground: it is the weight of the envelope and of the hydrogen it contains. At the beginning, the upward force was greater than the weight of the balloon, however, and now it is equal to the weight of the balloon. The upward force has therefore decreased.

Hence the upward or buoyant force of the air decreases with increasing altitude. It is equal to the weight of the volume of displaced air. This volume has not decreased, however. We have seen that the volume of a balloon tends rather to increase when the external air pressure decreases. What has obviously decreased is the weight of a quart of atmospheric air.

At sea level, 1 liter (1 quart) of air weighs 1.3 grams (.05 ounces). But at 1000 meters (3280 feet) altitude, it weighs less than 1.2 grams (.04 ounce). At 3000 meters (9840 feet) it weighs less than 1 gram (.035 ounce), and so forth. When the air pressure decreases, the weight of a liter of air also decreases.

This is not astonishing if one thinks about it. When a tire is inflated, more air is pumped into it. The greater the air pressure in the tire, the greater the mass of air in it. The volume of the tire practically does not change. Therefore, the mass of each liter of air has increased at the same time as the pressure: 1 liter (1 quart) of air at 2 atm is heavier than 1 liter (1 quart) of air at 1 atm (and at the same temperature).

What is strange, however, is that the situation is completely different with water. The mass of 1 liter (1 quart) of water is practically the same at the surface and at 5000 meters (16,400 feet) depth, where a pressure of about 500 atm exists!

Therefore, the buoyant force on a body submerged in water is always the same when the volume of the body remains unchanged. In this case, a sinking body does not stop at a certain depth. If it sinks below the surface, it continues to sink until it reaches the bottom. If it starts rising from the bottom, it continues to rise until it reaches the surface.

Molecules

When a bottle is full of water, it is impossible to put anymore in it. Well, what a profound statement! Listen! The fact is not as obvious as it seems at first sight. For instance, if a bus is full of people, it is still possible to squeeze in a few more.

And when a bottle is full of air?

L et us imagine an open air bottle on the beach. It is full of air at atmospheric pressure. If it has a volume of 7 liters (6.3 quarts), it contains about 10 grams (.35 ounce) air. The bottle is then hooked up to a compressor that will pump in more air; therefore the pressure in the bottle will increase.

When it reaches 10 atm, the bottle holds about 100 grams (3.5 ounces) air. Is it full? Yes, it is full of air at a pressure of 10 atm. However, if the compressor keeps operating, it can pump still more air into the bottle. At 20 atm, the bottle holds about 200 grams (7 ounces) air.

This process can go on. One stops in general when the air in the bottle is at 200 atm. At that time, the bottle holds about 2 kilograms (4.4 pounds) air! If one stops there, it is done as a precaution so that the bottle does not explode. In fact, one could continue—with a sufficiently powerful compressor—and into this bottle full of air at 200 atm one could still pump more air!

In contrast, a bottle of water at 200 atm would hold practically the same quantity of water that it would at 1 atm. If the bottle has a volume of 7 liters

(6.3 quarts), it holds 7 kilograms (15.4 pounds) of water at 1 atm (and at normal temperature). At 200 atm the bottle would hardly hold an additional few grams. Disregarding this small difference, one can say that the bottle still holds 7 kilograms (15.4 pounds) of water whatever the pressure of the water.

There is thus a huge difference between gases and liquids. It is said that gases are "compressible" (with increased pressure, more gases can be compressed in a bottle) whereas liquids are practically "incompressible." What is the reason for the difference between them?

Molecules

Water is formed by small solid particles, all the same, which are called *water molecules*. A molecule is much too small to be visible: a glass of water holds millions of billions of billions of them! Even though water is formed by these small particles, it is virtually impossible to see them, even with the most pow-

erful microscopes. Similarly, when flying over the Sahara desert at 5000 meters (16,400 feet) altitude, one cannot see that the sand consists of grains.

If water is cooled, it changes into ice. However, the molecules themselves do not change at all. The only difference is that in ice, they are attached one to the other to form a kind of sturdy scaffolding, whereas in water they are not attached: they move in all directions, remaining, however, pressed against one another.

When laundry dries, the water it holds "evaporates;" that is, it changes into vapor. Water vapor is a gas that mixes with air. Nonetheless, the molecules remain unchanged! However,

this time they no longer stick together; they are separated one from the other and each molecule moves in its own direction among air molecules.

Indeed, air is also formed of molecules. These molecules are not the same as those of water. Air molecules are of various kinds, mainly nitrogen and oxygen. Since air is a gas, these molecules are neither linked together the way they are in ice, nor are they pressed against one another the way they are in water. They are free and move in all directions, hitting and bouncing off each other endlessly.

In the normal air surrounding us, molecules travel at an average speed of 500 meters/second (1640 feet/second), and each one is hit about a billion times per second! However, between two impacts, a molecule travels all alone in a straight line and does not attach itself to other molecules.

Hence, this is the difference between a liquid and a gas: molecules of liquids are closely pressed together whereas those of gas are free and have much space in which to move about.

Because there is much space left between gas molecules, it is possible to add more to the same volume.

In contrast, molecules of liquids are already so close together that there is not much space left for additional molecules.

The Pressure of a Gas

The pressure of a liquid against the wall of a bottle is quite easy to understand. Molecules are not only pressed against each other, but also against the wall: they press against it the way that people in a crowded subway lean against the walls (and each other).

The pressure of a gas is more difficult to understand because molecules are not crowded at all. They travel in all directions, bouncing off each other and also off the walls. When they hit the wall and bounce back, they put pressure on the wall the way a tennis ball hits a racket. The impact is of course very short and the ball—or the molecule—does not press for very long. Nevertheless, the number of molecules is such that a small portion of the wall receives billions and billions of impacts in one second and the total effect is that of a continuous force.

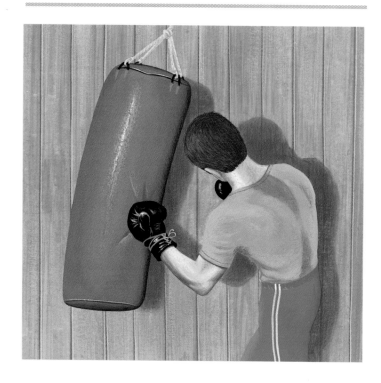

To understand better, once can imagine a boxer who hits a punching ball, that is, a bag full of sand hanging on a rope. If the boxer hits the bag hard, it moves away and then returns toward the boxer.

However, if the boxer gives many small blows, the bag does not have time to return: it remains pushed back, its rope being inclined exactly as if somebody pushed it sideways with a continuous force. The rapid bombardment of small blows keeps the bag there the same way a continuous force would.

However, not even a champion could hit the bag more than four or five times per second, whereas the wall that surrounds the gas receives billions of billions of impacts per square inch every second. This ultrarapid bombardment really has the effect of a continuous force.

What causes this force? The strength of the impacts and their number per second. The force thus certainly depends upon the number of molecules per quart and the movement of these molecules. Their movement in turn depends upon temperature: the higher the temperature, the faster the movement of molecules.

In short, the force per square inch of the wall, that is, the pressure, depends only upon two things: the temperature of the gas and the number of molecules per quart. At the same temperature, the pressure is twice as high if twice as many molecules occur per quart, three times greater if three times more molecules occur per quart, and so forth.

The Tire and the Pump

What keeps a tire inflated is the difference in pressure between inside and outside. If one presses with the thumb on a tire to check if it is well inflated, it is the difference in pressure between the inside and outside of the tire that prevents the thumb from pushing it in.

It is also this difference in pressure that enables the valve of the inner tube to prevent the air from escaping the tire. In fact, this valve consists of a small disk attached to the stem, which sticks out and sits inside the opening. For air to escape, this disk must be loosened from the opening. It is kept in that position by the difference in pressure between the inner and

When the pressure inside the tire is greater than the outside pressure, the disk is well seated against the edges of the opening.

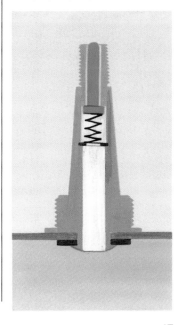

outer air. To loosen it, this difference in pressure must be overcome, either by pushing down the top of the stem with a finger or by pumping.

Let us imagine, for instance, that a tire is already well inflated and that additional pumping is done. The handle of the pump is pulled all the way. The pump is full of air at atmospheric pressure (we shall soon see how it entered). The air of the tire, being at a higher pressure, has its valve well shut.

If we start pushing in the handle of the pump very gently, we feel a resistance, which increases. This means that the pressure of the air locked up in the pump increases the more we push the handle. Why?

When pushing in the handle, we decrease the space occupied by the air in the pump. This is still the same amount of air; that is, the number of molecules is the same. If the same number of molecules occupies a smaller space, *the number of molecules per quart* increases. Consequently, the pressure also increases since it depends only upon the number of molecules per quart (and upon temperature, which is unchanged).

In other words, when a volume occupied by a given mass of air is decreased (without temperature change), the pressure of this gas increases.

The more we push in the handle of the pump, the higher the pressure of the enclosed air (and we have to push harder and harder). At a certain moment, this pressure is greater than that in the tire. Now the disk of the valve separates from the opening: the air of the pump communicates with the air in the tire.

Thereafter, when pushing in the handle, we decrease the volume occupied by all the air, in both the tire and the pump. Pressure continues to increase but not as rapidly as before. At the same time, air flows from the pump to the tire. When the handle is pushed in all the way, little air is left in the pump. The pressure in the tire has increased, and in the same way so has the number of air molecules in the tire.

As soon as one starts pulling the handle, the pressure in the pump decreases and the valve closes immediately: the air cannot escape from the tire. Now the pump must be filled with atmospheric pressure again. How?

To understand, let us dismantle a pump. We observe that a piston mounted on the stem of the handle and sliding in the tube is a disk of plastic in the shape of a cup whose convex side is turned against the handle.

When the pump is full of air and when the handle is pushed, the pressure of this air becomes greater than the atmospheric pressure: the air presses harder against the concave side of the disk and strongly presses its edges against the tube: air cannot escape from the pump by going around the disk.

When the handle is being pulled, however, the pressure inside the pump decreases more rapidly than the pressure of the outside air. This time, air presses more strongly against the convex side of the disk: these forces have a tendency to separate the edges of the disk from the tube.

Air can flow through on the sides and the pump is filled with atmospheric air until the handle is pushed again. At that time, the edges of the disk press anew against the wall of the tube with a strength increasing with the pressure in the pump.

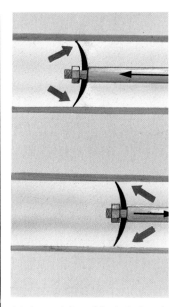

It is the curved shape of the plastic piston that allows the piston to press against the wall when the pump is pushed and to become separated from it when the pump is pulled.

Thus, when the volume occupied by a given mass of gas is decreased, without temperature change, the pressure of this gas increases. If, on the other hand, the volume is increased, the pressure decreases.

This explains the behavior of balloons in the atmosphere. However, once more, let us first see what happens to a "balloon" in the sea.

Diving

The dream of exploring the depths of the sea is almost as old as humans themselves. Did we hope to find natural wonders or retrieve wrecks of great value? Or was it simple curiosity about a world so close and yet so inaccessible? At any rate, more than 2000 years ago, the oldest of all instruments allowing people to breath under water was already in use: the diving bell.

Humans need to breathe in order to live. If they want to stay under water longer than some tens of seconds, they must carry their supply of air or arrange that it be sent from the surface.

The Diving Bell

The first diving bells date back to antiquity. Alexander the Great is often said to have used one himself to dive to the bottom of the sea during the siege of the port of Tyre in 322 B.C.

These ancient bells were merely heavy containers, open at the bottom and suspended from ropes. When the bell was above water, it was full of atmospheric air. The diver sat there and the dive began.

While the bell sank, the pressure of water increased and water therefore pressed increasingly upon the air enclosed in the bell. Water thus had a tendency to rise into the bell, thus decreasing the volume occupied by air. Therefore, the pres-

sure of this air increased so that it always remained equal to that of water at the level of its surface in the bell.

Suppose, for instance, that this surface is at 10 meters (33 feet) depth. It is known that the pressure of water at that depth is 2 atm. The pressure of the air enclosed in the bell is thus also 2 atm. It has doubled since the beginning of the dive. As learned in the last chapter, this means that the number of molecules per quart has doubled, too. The number of enclosed air molecules, however, has not changed, but the volume occupied by this air is therefore two times smaller than at the beginning. The water now fills the lower half of the bell, whereas the upper half is filled by air at 2 atm.

The same experiment can be made on a smaller scale with a glass in a bathtub. However, since the total depth of a bathtub is no greater than 0.5 meter (19 inches), the pressure is only 1.05 atm and the amount of enclosed air decreases very little. Water does not rise much

The diving bell was, until the last century, the only means for people to live long enough under water to be able to work.

inside the glass. A better result is achieved with a bottle. Since it has a narrow neck, the same variation in volume gives a greater change in the water level inside the bottle.

In short, the pressure of the air enclosed in the diving bell remains equal to that of the water at the level of the inside surface. The ears of the diver feel the same sensations as if he swam in water at the same depth.

If air is sent to the bell from the surface with a pipe, this air must be at least at the pressure that exists in the bell. Otherwise the air of the bell would escape through the pipe! However, if air is pumped with sufficient pressure, the bell can be filled and air can even be renewed while the surplus escapes from underneath the rim of the bell in streams of bubbles that rise to the surface.

Bubbles

Let us imagine a well-rounded bubble of air at a depth of 10 meters (33 feet). The surrounding water is at 2 atm. The air in the bubble has a slightly lower pressure because it is held by the surface of the bubble, which behaves like a stretched elastic "skin." The difference is small, however, so that it can be assumed that air in the bubble is at 2 atm, as is the surrounding water.

Of course, the bubble is much lighter than a water bubble of the same size and hence its weight is much smaller than the buoyant force: the bubble rises.

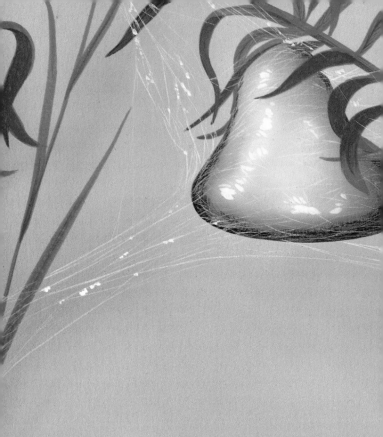

The Eurasian Aquatic Spider and Its Diving Bell

Argyroneta aquatica is an entirely aquatic spider that lives in ponds and slow-flowing rivers. This spider must breathe air and would be compelled to rise very often to the surface if it did not use a diving bell.

This bell has a web shape, like a bag open at the bottom, and is attached by numerous threads to vegetation at the bottom of the pond. To fill this bell with air, the spider climbs to the surface and exposes its belly to the air. Its belly is covered with hair that repels water. When the spider plunges into the water, its hair carries along a large bubble of air that resembles a silver armor. After entering its bell, the spider rubs its belly and the air bubble is trapped at the

top of the bag. Thereafter, the spider repeats this procedure as often as necessary to fill the bell with air. It can thus live there and lay eggs if the air inside the bell is periodically renewed.

During the bubble's ascent, it crosses successive zones where the water pressure becomes lower and lower. Hence, the pressure of the air in the bubble also decreases without change in the amount of air (or its temperature). Therefore, the number of molecules per quart decreases whereas the total number of molecules remains unchanged. Hence, the volume of the bubble increases and keeps increasing during its ascent.

The Cartesian Diver

This is an appartatus that can be easily built. To start with, one needs a rather tall glass container (such as a large jar for pickles or a container for orange juice), as well as an elastic rubber membrane, such as a piece of an inflatable toy balloon.

The Cartesian diver itself is a hollow small tube closed at one end. If it is transparent, so much the better. Nevertheless, this is not absolutely necessary. To the bottom of this tube is attached a

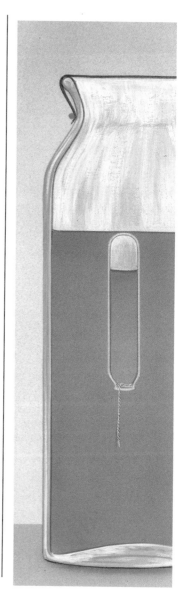

weight that compels the tube to float vertically with its opening toward the bottom.

Very slowly, water is let in (by tilting the tube) until it barely floats. It must emerge just a little above the surface of the water.

The glass container is filled almost completely with water, and the Cartesian diver is placed inside so that it just barely floats. The container is then closed with a rubber membrane that is well stretched and held tightly around the glass container with rubber bands.

If the rubber membrane is pushed down with the hand, the Cartesian diver sinks very slowly to the bottom. If the pressure of the hand is released, it rises again very slowly. With some training, it is possible to have it rise before it reaches the bottom and sink before it reaches the surface. However, it is practically impossible to have it stay still halfway up.

This behavior is easy to explain. When one presses on the rubber membrane, the air enclosed above the water in the container is compressed. The pressure of this air, and hence also that of water upon which it rests, increases.

Therefore, the pressure of the air enclosed in the Cartesian diver also increases, and as in the case of the air enclosed in the diving bell, its volume decreases. The buoyant force against the Cartesian diver thus decreases. As this force is only a little greater than the weight, it decreases faster than the weight itself: the Cartesian diver sinks.

The more it sinks, the higher the water column above it, and hence also the pressure it undergoes: the volume of air enclosed in the Cartesian diver decreases more and more, as does the buoyant force. The Cartesian diver is thus less and less supported.

When the hand is removed the pressure above the water again becomes equal to the atmospheric pressure, and the pressure at the bottom of the water decreases, too. The volume of air in the Cartesian diver increases again, and (if the water column above it is not too high) the buoyant force is again greater than the weight. The Cartesian diver rises, its volume increases more and more, and its tendency to rise increases until it reaches the surface.

This is the reason it is practically impossible to keep the Cartesian diver still "between two waters." Indeed, if one presses a little more, it sinks and descends faster and faster. If one presses a little less, it starts rising and does so more and more rapidly! This is the so-called unstable equilibrium, similar to that of a ruler held vertically on one finger. If it starts to deviate from the vertical line, it tends to do it more and more.

Fish and Submarines

Both behave somewhat like the Cartesian diver: they contain air whose volume decreases when they dive. The fish holds the air in its "swim bladder," the submarine in tanks that are more or less filled with water. Of course, the first can inflate its swim bladder a little more or a little less. Similarly, a submarine can pump more air into its tanks or let in more water. Both could thus remain at a more or less constant depth.

However, these adjustments would have to be done constantly because their equilibrium, as that of the Cartesian diver, is unstable. Hence, both prefer to use another method.

It is quite simple for the fish:

it gently jerks its flippers. The submarine moves forward while tilting at a variable angle small "wings" that play the same role as those of planes, but at a much smaller scale. The wings do not have to support the weight of the submarine but merely the difference (some tens of pounds) between the buoyant force and the weight of the submarine. For security reasons, one is in general careful to have the buoyant force a little greater than the weight. It is thus necessary to tilt the wings to that they pull down the submarine. If there is a breakdown, the submarine stops. Its wings are of no use when the submarine is no longer moving. Hence, the buoyant force brings the ship to the surface.

In short, if the submarine is not moving ahead, it cannot stay between two waters: it either floats or rests on the bottom.

Modern nuclear submarines finally made Jules Verne's dream come true: to travel "thousands of leagues" under water and even beneath ice caps. Indeed, their engines do not need any air compared to diesel engines, which operate only at the surface.

A Diver's Lungs

When breathing, we must inflate the chest to fill the lungs with air. In open air, there is no problem: the pressure of air that we breath is the same as that pressing upon our bodies. However, if a diver tries to breathe atmospheric air at a depth of 10 meters (33 feet) by means of a pipe reaching the surface, the situation is quite different.

The pressure of water at 10 meters (33 feet) is 2 atm. With air at 1 atm in the lungs, the difference in pressure between the outside and the inside air is 1 atm, that is, 1 kilogram/square centimeter (2.2 pounds/.155 square inch). The diver would have as much difficulty breathing as if he were stretched out on the beach with a weight of more than 1 ton on his chest!

A diver is therefore unable to breath atmospheric air. He must breath compressed air at the same pressure as that surrounding him: 2 atm at 10 meters (33 feet) depth, 3 atm at 20 meters (65.6 feet), and so on.

However, there is one inconvenience in breathing compressed air. After 2 atm, nitrogen, one of the constituents of air, starts to dissolve in the blood, which spreads it all over the body. The amount of nitrogen thus dissolved increases with pressure, that is with depth, and with the length of diving.

If the diver returns rapidly to the surface after a deep and long dive, nitrogen bubbles can form in the diver's blood system and cause extremely severe and even fatal blockages.

To avoid these dangers, a decrease in pressure during ascent must occur progressively. A diver who has remained half an hour at 30 meters (98 feet) depth must rest at least 5 minutes at 10 meters (33 feet) and 5 minutes at 3 meters (9.8 feet). If he dives even deeper or longer, these stops, called "decompression steps," must be more frequent and longer.

Such decompression steps can be avoided with the use of a "decompression chamber." This is a kind of airtight cabin fed by air at whatever pressure is desired. This chamber is low-

Even if he receives air from the surface through a pipe, the diver breathes compressed air at the same pressure as the water surrounding him: at 10 meters (33 feet) depth, the air is at 2 atm.

ered by a cable and the diver enters it, closing the door. The water inside is displaced by air at a convenient pressure. The diver thus continues breathing air at high pressure while the chamber is lifted onto the bridge of the ship. The pressure of air sent inside the chamber is then reduced very slowly. After very long and deep dives, it may take several hours before the pressure has returned to 1 atm and before the diver may finally leave the chamber.

The Diving Sperm Whale

A sperm whale is able to dive to a depth of 1000 meters (3280 feet), where it remains a quarter of an hour and then returns to the surface without going through "decompression steps." How is this possible?

First, a sperm whale breathes in deeply for several minutes at the surface until its blood is very rich in oxygen. Second, during its dive, its heart slows, which allows its oxygen reserve to last longer.

Third, it can do without decompression steps because it is not hooked to air bottles! In contrast to the diver, *it does not breathe compressed air*. The only amount of nitrogen to which its blood is exposed is that contained in its own lungs since the beginning of its dive. This amount is not sufficient for the dissolution of nitrogen in blood to become dangerous, however.

Furthermore, the sperm whale's sturdy body frame offers resistance to pressures of more than 100 atm!

And why is it diving to such depths? To find its favorite food there: giant squid. These animals live only under very high pressure and never rise to the surface. Nobody has ever seen them. We know about their existence because sections of giant tentacles with suckers as large as dishes have been found in the stomachs of sperm whales.

The smallest species of the family of whales are even more agile then sperm whales. However, they do not dive as deeply as sperm whales.

To Rise in the Air

When Archimedes discovered the principle of buoyancy, people had already been using the buoyant force of water for several thousands of years: they swam and navigated without asking themselves why they could do it.

However, to use the buoyant force of air to "navigate in the atmosphere," it was first necessary to believe that it might be possible to do so. In this case, investigation preceded development. Based on Archimedes' discovery 2000 years earlier, people invented the means to rise in the air.

The Invention of the Brothers Montgolfier

Paper manufacturers at Annonay, in Ardèche (France), in the years 1780, the two brothers Etienne and Joseph Montgolfier were fascinated by the discoveries about the behavior of gases at that time.

Soon after its publication, they read the book by the English naturalist Joseph Priestley, *Experiments and Observations on Different Kinds of Air* (which means "on different gases"). In this book, Priestley wrote about the properties of various gases discovered by others and by himself and showed, in particular, that some gases are lighter than air.

Upon reading this, the brothers Montgolfier thought that when filling a very thin envelope with light gas, one could make an object that is lighter than air. The buoyant force would be greater than the weight of the object, and the latter would rise in the air.

The first human flight occurred in 1783 when Pilâtre and d'Arlandes flew over Paris in a hot-air balloon (montgolfière in French).

After several trials, the brothers filled an envelope of light fabric, open at the bottom, with "something resembling a cloud": the smoke of burning humid straw mixed with strings of wool. The contraption took off!

As we shall soon see, it was neither the smoke nor the humidity that made the contraption rise: it was hot air. From then on it was called a montgolfière (hot-air balloon). To the brothers Montgolfier, however, it did not matter! They had invented the first airship capable of taking off!

On June 4, 1783, they made a public demonstration in their city of Annonay with a large hot-air balloon, 12 meters (39 feet) in diameter, made with fabric lined inside with paper. Underneath the opening was attached a wire basket that carried the burning straw. In ten minutes, the balloon rose to an elevation of 500 meters (1640 feet).

News of this successful event soon reached Paris, and the Academy of Sciences invited the brothers Montgolfier to run a demonstration of their balloon in Paris, with all expenses

The first flying passengers, a lamb, a rooster, and a duck, were locked up in a wicker basket attached to one of the first hot-air balloons. Everything went well for them: the sky was open to humans!

paid by the Academy.

This time, the hot-air balloon carried passengers: a lamb, a rooster, and a duck riding in a wicker basket attached to the balloon. On September 19, 1783, in the presence of the King Louis XVI and a huge crowd in the large courtyard of the castle of Versailles, the hot-air balloon took off with its "passengers." It fell to the ground some miles away, but the animals did not seem to have suffered from their trip. The sky was now open to humans!

In an even larger hot-air balloon, Pilâtre de Rozier and the Marquis d'Arlandes flew over Paris two months later from a place called "la Muette" to another one, "la Butte aux Cailles" (the hill of the quails). The human conquest of the sky had started.

Why Does a Hot-Air Balloon Rise?

The "gas resembling a cloud" with which the brothers Montgolfier filled their balloon was simply hot air. That was enough to make it rise.

Indeed, we have learned that the pressure of a gas depends upon two things: the number of molecules per quart and temperature. To obtain the same pressure at a higher temperature, fewer molecules per quart are needed.

A hot-air balloon being open (at the bottom), the air pressure inside and outside is the same. However, the inside air is warmer and therfore holds fewer molecules per quart: *1 quart of this air is lighter than 1 quart of cold air that surrounds the balloon.* The weight of the inside air is thus lower than the weight of the displaced air. If the difference is greater than the weight of the balloon envelope itself, including the gondola and the passengers, the hot-air balloon rises.

However, it does not rise forever. Having reached a certain altitude, it stops as does a balloon inflated with hydrogen mentioned earlier. For a balloon to maintain this altitude, the fire must be kept going because if the inside air cools off, the balloon descends. This is a serious drawback that greatly reduces the potential of a hot-air balloon. A few weeks after its invention, it had a competitor that was soon to replace it: the hydrogen balloon.

The Two Balloons of Professor Charles

When the news about the invention by the brothers Montgolfier reached Paris, the general public was too impatient to wait for the two brothers to come to the capital for a demonstration. In a few days, an important sum of money was collected by subscription and the physicist J.A.C. Charles was put in charge of the immediate construction of a balloon that could rise in the air.

However, the brothers Montgolfier did not reveal their secret. They merely announced that they had filled their balloon with "gas twice as light as ordinary air." Which gas? Charles had no idea. He finally decided to fill his balloon with an even lighter gas, fourteen times lighter than air: hydrogen.

Hydrogen had just been discovered, and it was still difficult to produce in large quantities. Furthermore, it escapes easily through most fabrics and is very flammable. Despite all these difficulties, Charles succeeded in filling with hydrogen a balloon made with taffeta, coated with rubber, and having a diameter of about 3 meters (9.8 feet).

On August 27, 1783, almost a month before the ascent of the hot-air balloon at Versailles with its "animal crew," Charles' balloon was carried by night, under strong security, to the Champ de Mars. At 5:00 p.m., in front of 300,000 persons (half the population of Paris!), the firing of a cannon announced its departure. The balloon was released and it rose immediately to an altitude of 1000 meters (3280 feet) before disappearing in the clouds.

The event lasted only two minutes, but the enthusiasm of the crowd was extraordinary: people hugged each other, laughed, or cried! A heavy rain began to pour, but everybody remained there staring at the place in the sky where the first hydrogen balloon had disappeared. It fell to the ground a little later in the suburbs of Paris.

The second had a longer career. Charles opened another subscription to build a hydrogen balloon that could carry two passengers. In a few days,

In the past few years, hot-air balloons have become fashionable again in the form of a sport, which is rather expensive. These modern hot-air balloons, made with ultralight cloth, carry a propane burner and one or two passengers.

the money was collected and the construction began.

This time, the balloon had a diameter of 9 meters (29.5 feet) and was provided with astonishing improvements: a more sophisticated rubberized fabric; a net supporting the gondola; ballast that allowed a fast rise; a safety valve that released gas for a gentler descent; and a barometer to measure the altitude reached! All of a sudden, the balloon had grown up!

On December 1, 1783, ten days after the ascent by Pilâtre, Charles and his assistant M.N. Robert took off from the Tuileries. They traveled some 40 kilometers (25 miles) before landing. Robert left the gondola and thus the lightened balloon was able to rise again fast. In ten minutes, Charles reached an altitude of 4000 meters (13,125 feet). After a few measurements, he opened the valve and started to descend. He landed half an hour later and some miles farther away. Less than six months after the first experiment by the brothers

Some sports fans still use balloons that very much resemble the one made by Charles but are inflated with helium. This gas is a little heavier and much more expensive than hydrogen. However, it is not flammable!

Montgolfier, balloons had become truly "operational."

Following these events and for over a century, people tried to perfect these "dirigibles" until the first planes brought a completely different approach to flying in the air. Thereafter, balloons were no longer used for transporation, except for pleasure trips. Nevertheless, they are still in use for other purposes because they can do one thing better than planes: rise very high.

To Rise Very High

At high altitudes, air is very rarefied, its pressure is very low, and 1 quart weighs almost nothing. Of course, 1 quart of hydrogen at the same pressure still weighs fourteen times less, and the difference between the two is still thirteen-fourteenths of the weight of a quart of air. However, the thirteen-fourteenths of almost nothing cannot amount to much. To carry along a cargo, the balloon must be gigantic.

On the other hand, for hydrogen to be at the same low pressure as air at high altitude, very little is required upon departure. Thus at this point, the

balloon is very limp and looks like an almost empty envelope. With increasing altitude, the inside pressure decreases at the same time as the outside pressure and the balloon inflates little by little until it is well rounded and tight.

These pictures show the balloon (before departure, left) in which Professor Piccard rose to 16,900 meters (55,445 feet) on August 18, 1932. Once this envelope became inflated—at high altitude—it changed into a sphere measuring 30 meters (98 feet) in diameter with a volume of 14,000 cubic meters (18,310 cubic yards)! Indeed, the balloon had to carry, besides its own envelope, Professor Piccard and his assistant in an airtight and pressurized cabin: otherwise they would not have reached such altitudes alive.

This cabin was a sphere of aluminum shown below at the time of departure before Professor Piccard had drawn in his head and closed the porthole. This is one of the last pictures of the history of balloons. It is also a picture that would have gladdened the heart of Galileo—the first to weigh air—and of course of Archimedes.

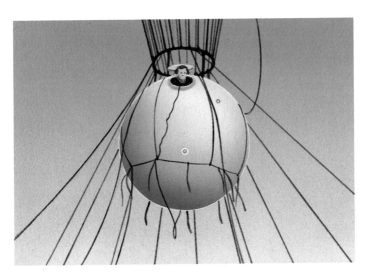

Fragile Monsters

Between 1850 and 1900, many inventors tried to render balloons "dirigible," that is, independent of wind. For that reason, they needed a slender aerodynamic shape, and above all an engine that could activate propellers. However, an engine and its fuel are very heavy. It was only in 1900 that it was possible to build both rather light engines and quite large balloons to support them. The result was the German *Zeppelin*.

This zeppelin was an enormous streamlined airship, made rigid by a metallic framework and divided into several pockets to render a possible leak less catastrophic. The largest—and last—zeppelin, called the "Hindenburg," was 247 meters (810 feet) long and 45 meters (147 feet) high and weighed 195 tons. It crossed the Atlantic several times with a crew of 40 and 50 passengers. On May 6, 1937, while landing at Lakehurst, New Jersey, the hydrogen-inflated zepplin, the "Hindenburg," burst into flames and was completely destroyed. The tragedy resulted in the loss of 36 lives.

At any rate, at that time, the airship already had a more and more serious competitor: the airplane.

Index

The numbers in italics refer to illustrations.

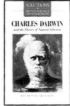

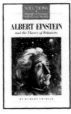